INTRODUCTION

The practical investigations in this *Guide* relate largely to the topics covered in *Study guide I*, Part Two, 'Control and co-ordination in organisms', Chapters 11 to 14. Cross references to the *Study guide* are given.

Chapter 11 CO-ORDINATION AND COMMUNICATION

Investigation 11A The spinal cord and spinal roots. (*Study guide* 11.2 'The monosynaptic reflex'.)
Microscopical preparations are used to study the anatomy and histology of the spinal cord and the roots of spinal nerves.

Investigation 11B The histology of nerve and muscle tissue. (*Study guide* 11.3 'The neurone' and 11.6 'Skeletal muscle'.)
Microscopical preparations are used to study the histology of nerve and muscle tissue.

Investigation 11C The physiology of a nerve–muscle preparation. (*Study guide* 11.4 'The nerve impulse' 11.5 'Chemical messages', and 11.6 'Skeletal muscle'.)
An investigation into the physiology of nerves and muscles by stimulating the nerve and recording the contraction of the muscle on a kymograph.

Investigation 11D The nerve impulses in the earthworm's ventral nerve cord. (*Study guide* 11.4 'The nerve impulse'.)
A simple reflex arc is investigated by using a cathode-ray oscilloscope to record the nerve impulse.

Chapter 12 THE RESPONSE TO STIMULI

Investigation 12A The response of *Tribolium* to humidity. (*Study guide* 12.2 'Animal responses'.)
A choice chamber is used to investigate this response.

Investigation 12B The response of *Calliphora* larvae to light. (*Study guide* 12.2 'Animal responses'.)
The type of response which an animal may exhibit towards light is investigated.

Investigation 12C Short-term changes in responsiveness. (*Study guide* 12.2 'Animal responses'.)
The waning of an earthworm's responsiveness to the stimulus of touch is studied.

Investigation 12D The tropic responses of coleoptiles and radicles. (*Study guide* 12.3 'A plant's response to light' and 12.4 'A plant's response to gravity'.)
Phototropic and gravitropic responses are investigated in coleoptiles and radicles viewed down the microscope.

Investigation 12E The nature and detection of the stimulus involved in phototropism. (*Study guide* 12.3 'A plant's response to light'.)
The particular wavelengths of light which are important in a phototropic response are investigated.

Chapter 13 **BEHAVIOUR**

Investigation 13A Turning behaviour in woodlice. (*Study guide* 13.2 'Studying the causation of behaviour'.)
A causal explanation of the behaviour of woodlice in a maze is investigated.

Investigation 13B The reproductive behaviour of the three-spined stickleback. (*Study guide* 13.2 'Studying the causation of behaviour'.)
The behaviour of sticklebacks in an aquarium is observed. Other animals or filmed material can be used instead.

Investigation 13C Sex pheromones in the Mediterranean flour moth. (*Study guide* 13.2 'Studying the causation of behaviour'.)
The function of pheromones as sexual attractants is investigated.

Investigation 13D Associative learning in mammals. (*Study guide* 13.2 'Studying the causation of behaviour'.)
Small mammals are used to investigate the effect of previous experience on learning.

Investigation 13E Social order in hens. (*Study guide* 13.4 'Studying the functions of behaviour'.)
The observation of a small group of hens, at first hand or on film.

IDE 4

SPONSE,

NUFFIELD FOUNDATION. Nuffield Advanced Science
Biology Practical Guide 4.

Longman Group UK Limited
Longman House, Burnt Mill, Harlow, Essex CM20 2JE, England
and Associated Companies throughout the world

First published 1970
Revised edition first published 1985
Fifth impression 1992
Copyright © 1970, 1985, The Nuffield-Chelsea Curriculum Trust

Design and art direction by Ivan Dodd
Illustrations by Oxford Illustrators

Set in Times Roman and Univers
Produced by Longman Singapore Publishers Pte Ltd
Printed in Singapore

ISBN 0 582 35430 7

The publisher's policy is to use paper manufactured from sustainable forests.

Cover illustration
Motor neurones (×445). See investigation 11A, 'The spinal cord and
spinal roots'.
Photograph, Biophoto Associates.

CONTENTS

SAFETY

In these *Practical guides*, we have used the internationally accepted signs given below to show when you should pay special attention to safety.

 highly flammable

 explosive

 toxic

 corrosive

 radioactive

 take care! (general warning)

 risk of electric shock

 naked flames prohibited

 wear eye protection

 wear hand protection

Chapter 14   THE HUMAN BRAIN AND THE MIND

Investigation 14A Sensorimotor skills in humans. (*Study guide* 14.1 'The human brain'.)
An exercise involving mirror drawing and the effect of experience on the performance of a skill.

Investigation 14B Perception. (*Study guide* 14.2 'The mind and consciousness'.)
An examination of visual illusions.

Investigation 14C Memory. (*Study guide* 14.2 'The mind and consciousness': 'Memory and learning'.)
This investigation tests the recall of information by the brain.

Investigation 14D Intelligence. (*Study guide* 14.2 'The mind and consciousness': 'Intelligence'.)
In this investigation the skills which seem to be reliable indications of intelligence are tested.

Investigation 14E Emotion. (*Study guide* 14.2 'The mind and consciousness': 'Emotion and feeling'.)
Certain physical signs are used to indicate the state of arousal of the brain.

Investigation 14F Sleep. (*Study guide* 14.2 'The mind and consciousness': 'Sleep'.)
This investigation is an attempt to establish a pattern of night and day rhythms.

Investigation 14G Extra-sensory perception and psychokinesis. (*Study guide* 14.2 'The mind and consciousness': 'Changing the way we see the world'.)
Simple experiments on extra-sensory perception are carried out.

A note for users of this *Practical guide*

The instructions given for the investigations are intended for use as guidelines only. We hope that you will modify and extend the techniques that have been described to meet your own requirements. Other organisms should certainly be tried, depending on what is most readily available. Some of these investigations may lend themselves to further work in a project.

It may not always be possible, for various reasons, for you to do a practical investigation yourself. A study of data from another source is perfectly acceptable in such a case.

CHAPTER 11 **CO-ORDINATION AND COMMUNICATION**

INVESTIGATION
11A The spinal cord and spinal roots

(*Study guide* 11.2 'The monosynaptic reflex'.)

Nerve fibres enter and leave the spinal cord in the spinal nerves. These nerves lie in pairs on either side of the spinal cord and each nerve connects with the cord through two roots, one on the dorsal side and one on the ventral side. This investigation is concerned with the anatomy and histology of the spinal cord and the roots of the spinal nerves.

Procedure
You are going to examine a series of microscopical preparations. It is suggested that in each case you make appropriate drawings of the relevant features.

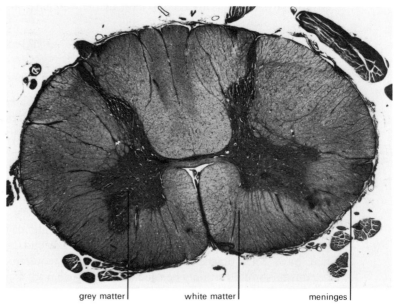

grey matter | white matter | meninges

Figure 1
A transverse section through the cervical region of a human spinal cord (× 7).
Photograph, Biophoto Associates.

1   Examine, under low power, a permanent microscope preparation of spinal cord (transverse section), if possible one in which the dorsal and ventral roots are included. Alternatively, view a 35 mm transparency.
2   Use *figure 1* to identify the white and grey matter and the meninges.
3   Examine the grey matter under high power. Find good examples of the cell bodies of the motor neurones and note their structure (*figure 2*).

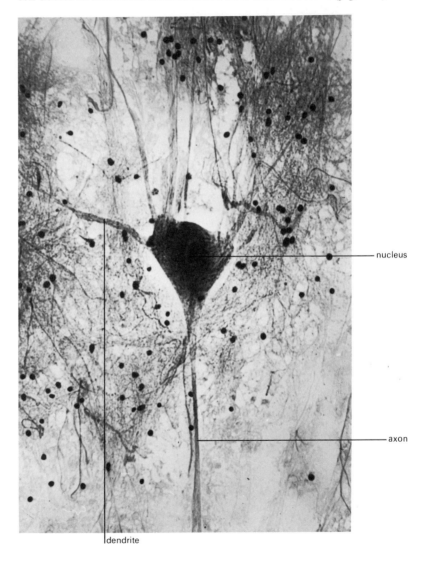

Figure 2
A motor neurone cell body from the spinal cord of an ox (× 233).
Photograph, Biophoto Associates.

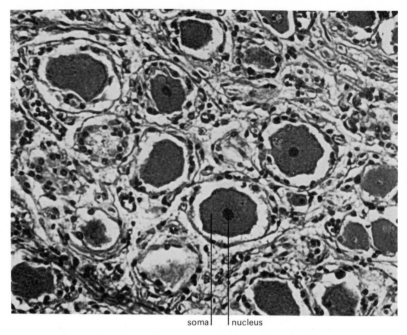

soma | nucleus

Figure 3
A longitudinal section through a dorsal root ganglion of a baboon (× 375).
Photograph, Biophoto Associates.

4 Examine the dorsal and ventral roots under high power. Identify the
dorsal root ganglion and the cell bodies of the sensory neurones that it
contains (*figure 3*). The ventral root consists entirely of nerve fibres.

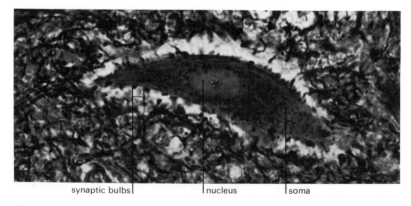

synaptic bulbs | nucleus | soma

Figure 4
A photomicrograph of a synapse on a neurone cell body in a transverse section of rabbit
spinal cord (× 1150).
Photograph, Biophoto Associates.

Note the components of the two roots.

5   *Figure 4* is a photomicrograph of synapses found in the spinal cord. *Figure 5* is an electronmicrograph showing the detailed structure of a synapse. Examine these micrographs, noting in particular the labelled structures.

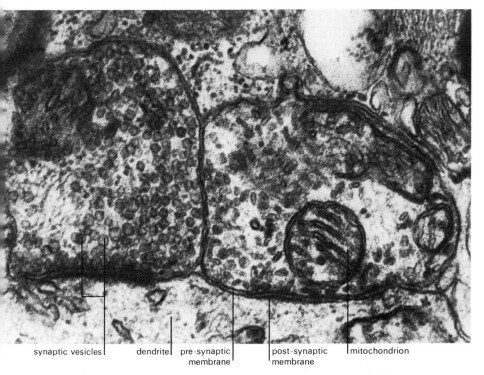

synaptic vesicles   dendrite pre-synaptic   post-synaptic   mitochondrion
                               membrane   membrane

Figure 5
An electronmicrograph of two synaptic bulbs on a dendrite from the spinal cord of a goldfish (× 55,000).
Photograph, Professor E. G. Gray, FRS.

Questions

a   *How can the dorsal root be distinguished from the ventral root by its gross appearance?*

b   *Where in the spinal cord and spinal roots are cell bodies found?*

c   *How can the cell bodies of motor neurones be distinguished from those of the sensory neurones?*

d   *In which region of the spinal cord would you expect synapses to be found?*

e *How do the arrangement of synapses and the anatomy of the cell bodies of the motor neurones relate to the function of the spinal cord in co-ordinating nervous activity?*

f *What features of the synapse's function can be inferred from figure 5?*

INVESTIGATION
11B The histology of nerve and muscle tissue

(*Study guide* 11.3 'The neurone' and 11.6 'Skeletal muscle'.)

The basic function of a nerve is to conduct information in the form of nerve impulses to and from the central nervous system. Muscles are closely associated with nerves and it is convenient to study the histology of both together.

Procedure
As in the previous investigation you are advised to make drawings of relevant features of each preparation examined.

1 Examine, under low power, a permanent microscope preparation of a nerve (transverse section). Identify the outer layer of connective tissue which surrounds bundles of nerve fibres (axons) held together by more connective tissue. Blood vessels may also be present. Note their arrangement.

2 Examine, under low and high power, a similar section that has been treated with osmium(VII) oxide (osmic acid) which stains fat black. Use *figure 6* to identify the Schwann cells and myelin sheaths that surround most of the axons. Note the appearance of myelinated axons as seen under high power.

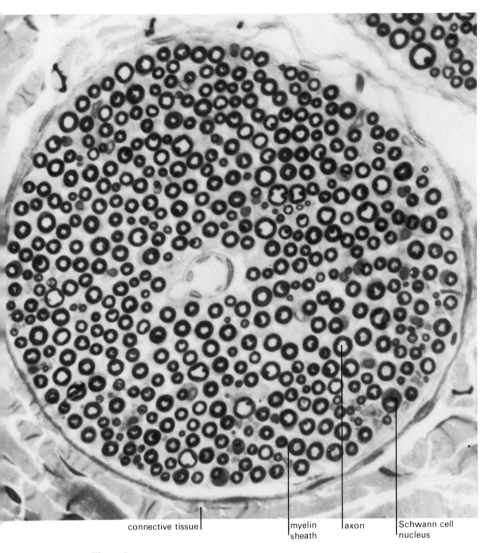

connective tissue   myelin   axon   Schwann cell
                             sheath   nucleus

Figure 6

A transverse section of part of the nerve of a rat ($\times$ 350).
Photograph, Biophoto Associates.

3 *Figure 7* is an electronmicrograph of a myelinated axon. Examine the micrograph, noting the labelled structures.

axon | myelin sheath

Schwann cell nucleus | cytoplasm Schwann

Figure 7
An electronmicrograph of a myelinated axon (× 43,000).
Photograph, Department of Human Anatomy, Oxford University.

4 Examine a longitudinal section of a nerve under low and high power. Find good examples of the nodes of Ranvier as shown in *figure 8*.

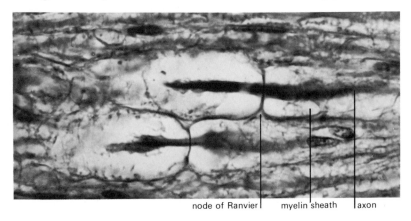

node of Ranvier | myelin sheath | axon

Figure 8
A longitudinal section of the posterior root nerve of a cat showing nodes of Ranvier
(× 750).
Photograph, Biophoto Associates.

5 *Figure 9* is an electronmicrograph of a node of Ranvier. Examine the node, noting how it relates to the myelin sheath.

node of Ranvier

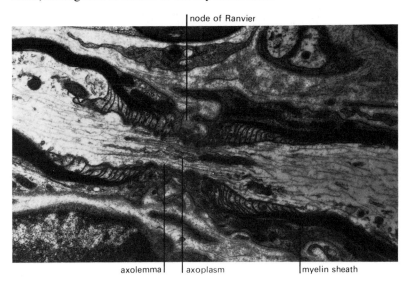

axolemma | axoplasm | myelin sheath

Figure 9
An electronmicrograph of a node of Ranvier (× 17 000).
Photograph, Dr A. R. Lieberman, Department of Anatomy and Embryology, University College London.

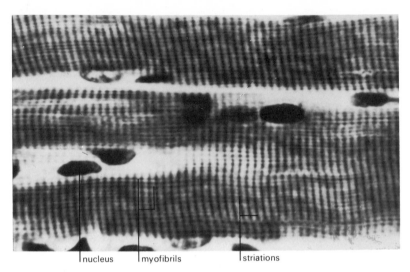

nucleus   myofibrils   striations

Figure 10
A photomicrograph of a longitudinal section of striated muscle from the rectus muscle of a monkey's eye (× 1140).
Photograph, Biophoto Associates.

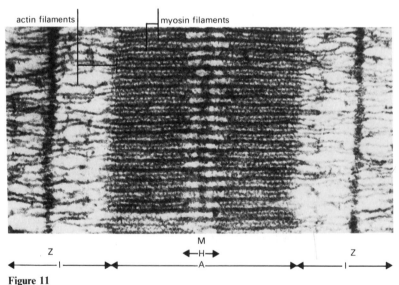

actin filaments   myosin filaments

Z   M ←H→   Z
I   A   I

Figure 11
An electronmicrograph of one sarcomere from a myofibril in rabbit muscle (× 82,000). The letters, A, H, I, M, and Z are the conventional ways of referring to the different parts of the myofibril.
Photograph, Dr H. E. Huxley.

6   Examine a transverse section of striated muscle under low power. Note the arrangement of muscle fibres, connective tissue, and blood vessels within the muscle.

7   Examine a longitudinal section of striated muscle under low and high power. Use *figure 10* to identify the nuclei and striations found on the muscle fibres. Note the appearance of the fibres as seen under high power.

8   *Figures 11* and *12* are electronmicrographs of the myofibrils contained within a muscle fibre. Observe the structure of the myofibrils and note how the filaments inside them are arranged.

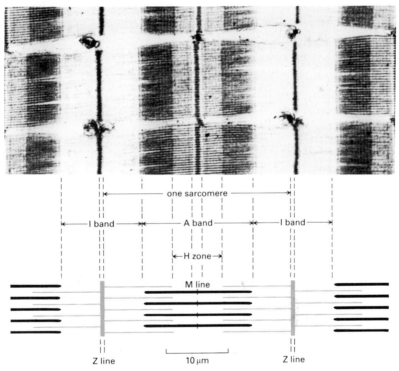

Figure 12
An electronmicrograph of a longitudinal section of frog muscle fibre showing parts of three myofibrils (× 20,200).
Photograph, Dr H. E. Huxley.

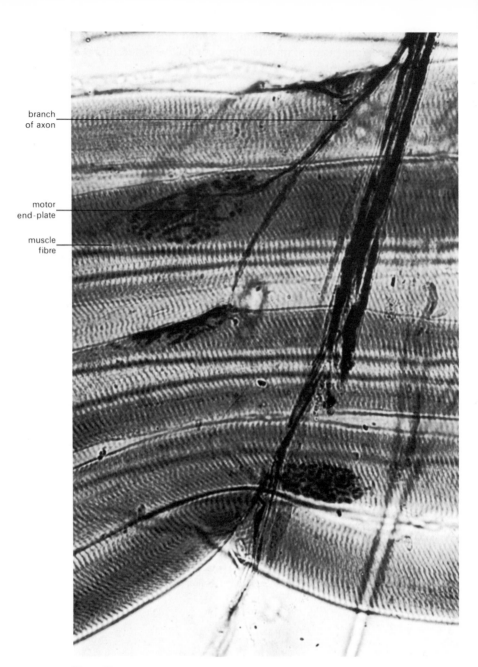

branch
of axon

motor
end-plate

muscle
fibre

Figure 13
Motor end-plates in snake muscle (× 750).
Photograph, Biophoto Associates.

9 Examine a permanent microscope preparation that shows the connections between the terminal branches of a motor axon and the muscle fibres which they innervate (*figure 13*).

10 *Figure 14* is an electronmicrograph of a motor end-plate. Note the relationship between the nerve terminal and the muscle fibril.

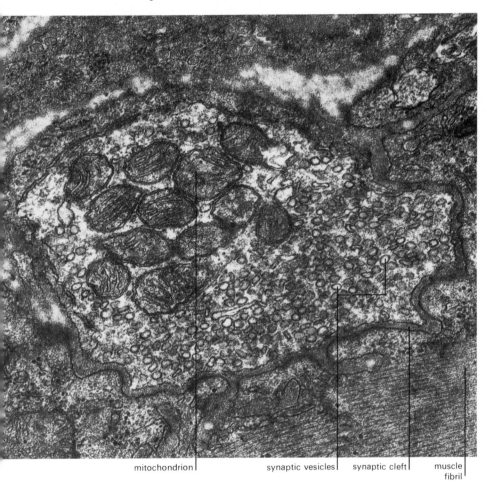

mitochondrion | synaptic vesicles | synaptic cleft | muscle fibril

Figure 14
An electronmicrograph of a motor end-plate in a mouse (× 43 000).
Photograph from Winlow, W. and Usherwood, P.N.R., 'Ultrastructural studies of normal and degenerating mouse neuromuscular junctions', J. Neurocytol. 4, 377–394, 1975.

Questions

a *What can you infer about the nature of the myelin sheath?*

b *What happens to the myelin sheath at the nodes of Ranvier?*

c *What processes could occur at the nodes of Ranvier which might be difficult or impossible at other parts of the axon?*

d *What is responsible for giving muscle fibres their striated appearance?*

e *Can you suggest a possible mechanism that could bring about the shortening of a muscle fibre?*

f *What is the relationship between a motor axon and the muscle fibres which it innervates?*

g *What features of the motor end-plate's function can you infer from figure 14?*

INVESTIGATION

11C The physiology of a nerve–muscle preparation

(*Study guide* 11.4 'The nerve impulse', 11.5 'Chemical messages', and 11.6 'Skeletal muscle'.)

Biologists have learnt a great deal about the physiology of nerves and muscles by using nerve–muscle preparations. The nerve of such a preparation can be stimulated by giving it an electric shock. Stimulators used for this purpose will normally provide electric shocks over a range of voltages and a variable number of repetitive shocks per second. The contraction of the muscle is best recorded by means of a kymograph. It is convenient to use a poikilothermic animal because it is not then necessary to maintain the animal's body temperature or continuously aerate the bathing liquid. The following instructions apply to the frog, which is the animal normally used for this kind of investigation.

Procedure

1 You are provided with a killed frog. Cut the skin around the middle of the body and, using fingers and blunt forceps, pull the skin backwards and off the hind legs. Keep the dissection moist at all times with Ringer's solution.

2 Identify the gastrocnemius (calf) muscle and the Achilles tendon in one leg. Using a seeker, separate the tendon from connective tissue and slip a piece of cotton thread behind the tendon. Tie the thread just anterior to the sesamoid bone as shown in *figure 15*.

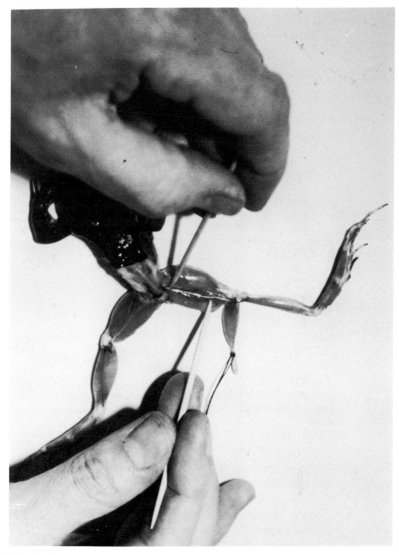

Figure 15
A dissected frog with gastrocnemius muscle and sciatic nerve exposed.
Photograph, Malcolm Fraser.

3   Cut the tendon posterior to the sesamoid bone and free the
gastrocnemius muscle from the tibio-fibula bone leaving the muscle
attached only at the knee joint.

4   Using wooden or plastic cocktail sticks, separate the muscles along the
back of the thigh and locate the sciatic nerve (*figure 15*). Carefully

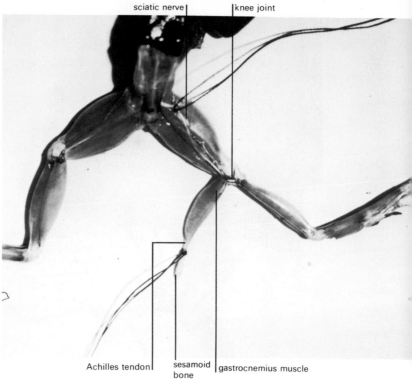

sciatic nerve | knee joint

Achilles tendon | sesamoid bone | gastrocnemius muscle

Figure 16
A hind leg of a frog with the gastrocnemius muscle deflected and the sciatic nerve exposed. Threads have been tied around the Achilles tendon and the upper end of the nerve.
Photograph, Malcolm Fraser.

trace the nerve as far back as you can and forward to the knee joint. At the anterior end slip another piece of cotton thread behind the nerve and make a knot. Your preparation should now look like *figure 16*.

5 Cut the nerve anterior to the knot and, still using cocktail sticks, carefully separate the nerve from the muscle as far as the knee joint.

6 Cut the femur 1 cm from the knee joint and the tibio-fibula just below the knee joint. Insert a pin through the knee capsule.

7 Place the preparation in a muscle bath containing Ringer's solution and push the pin into the cork base or the hole, if there is one. Tie the thread which is attached to the tendon to the lever and adjust the position of the lever until it is horizontal. Use the other thread to lift the nerve out of the solution and place it over the stimulating electrodes. Ensure that the nerve is kept permanently moist with Ringer's solution.

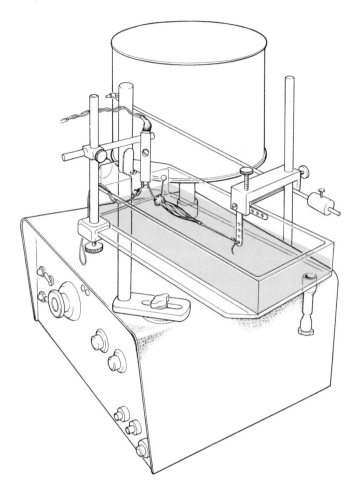

Figure 17
A muscle bath and kymograph.

8   Place a sheet of recording paper around the kymograph drum,
     ensuring that the edges overlap in the same direction as the recording
     pen moves. Check that the pen is writing smoothly and producing an
     even line. The complete arrangement is shown in *figure 17*.
9   Switch on the stimulator, set the voltage control at its lowest point and
     deliver one stimulus. Turn the drum by hand about 2 cm, increase the
     voltage slightly and repeat. Continue this until an increase in stimulus
     voltage gives no further increase in the amount of contraction. Remove
     the paper from the drum and below each trace record the size of the
     stimulus. Keep the stimulator at the optimum voltage for all
     subsequent experiments.

10   Arrange the switch on the kymograph so that one stimulus will be given for each complete drum rotation. Move the pen away from the drum and switch on the kymograph to run at maximum speed. Bring the pen up to the paper to produce an even line and operate the stimulator control to allow one stimulus to be delivered. Immediately the response has been made, switch off the kymograph.

11   To mark the position on the trace where the stimulus was delivered, turn the drum by hand until the kymograph switch is about to be closed. Operate the stimulator control and turn the drum slowly until the stimulus is delivered and the response recorded. Measure the distance between the stimulus mark and the beginning of the response and, from the time-base of the kymograph, calculate the delay period between a stimulus being delivered and the muscle responding.

12   Gently lift the nerve off the electrodes and lower it into the Ringer's solution. Arrange the electrodes on the surface of the muscle so that the muscle can be stimulated directly and yet be free to contract. Repeat the procedures to give the muscle a single stimulus and calculate the delay period as before.

13   Arrange the switching on the kymograph so that two stimuli can be delivered in succession. Set the kymograph to run at maximum speed and start with the two switches separated by about 45°. Record the response and repeat, each time bringing the two contact switches closer together until the effect is the same as that obtained by one switch alone. Calculate the time intervals between the successive stimuli in each case.

14   Set the kymograph to run at about one revolution per minute and arrange the switching to deliver repetitive stimuli. Record responses to different frequencies of stimulation up to 20 stimuli per second. You must allow the preparation a rest period of about 30 seconds between each set of stimuli. Measure the size of each set of recordings and compare these with the peak of a single twitch.

15   Finally, subject the preparation to continuous stimulation (20 stimuli per second) for 20 seconds and record the results.

Questions

a   *Why are wooden or plastic cocktail sticks used to separate the nerve from the muscle tissue?*

b   *During the course of the dissection, did the muscle show any signs of being stimulated? If so, what do you think caused the stimulation?*

c   *How do you account for the responses of the muscle to stimuli of different voltages?*

d *What is the period of delay between a stimulus being delivered to the nerve and the muscle responding? What mechanical and physiological factors could contribute to this?*

e *What is the period of delay when the muscle is stimulated directly? What can be deduced from this about the velocity of the nerve impulse?*

f *How could the nerve–muscle preparation be used to measure the velocity of the nerve impulse? State what measurements you would have to make.*

g *What effects do two successive stimuli have on the muscle?*

h *What effects do repetitive stimuli have on the muscle?*

i *From this investigation, what determines the amount that a muscle contracts?*

j *How do you account for the effect of continuous stimulation on the muscle?*

INVESTIGATION
11D The nerve impulses in the earthworm's ventral nerve cord

(*Study guide* 11.4 'The nerve impulse'.)

The ventral nerve cord of the earthworm contains several giant axons which transmit large impulses very rapidly along the length of the worm. These impulses are important in eliciting the worm's escape reflex in which the contraction of the longitudinal muscles quickly draws the animal back into its burrow in response to touch. A cathode-ray oscilloscope can be used to record the impulses and investigate some of the properties of this simple reflex arc.

The oscilloscope operates on the same principle as the cathode-ray tube of a television set. The end of the tube contains a filament which, when heated, emits a stream of electrons (the electron gun). These are focused to form a narrow beam that produces a spot of light when it strikes the fluorescent screen at the other end. The spot can be swept rapidly across the screen, from left to right, to produce a horizontal line of light. From the frequency of these sweeps the time base can be determined, that is, the time taken for the spot to travel 1 cm across the screen.

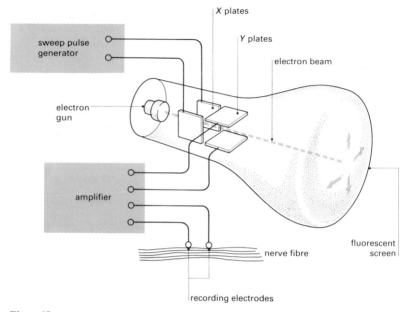

Figure 18
A cathode-ray oscilloscope arranged to measure nerve impulses.

Any potential difference to be measured is applied to the Y plates, which will deflect the spot up or down, the size of the deflection depending on the amount the signal is amplified. The arrangement is shown in *figure 18*.

The signal voltage from a nerve will appear on the oscilloscope screen as a momentary, stationary wave or action potential, as illustrated in *figure 19*. Because this has been amplified many times, the effects of unwanted electric fields, for example from the mains supply, must be eliminated by using coaxial cable and placing the preparation in an earthed metal dish.

Procedure

1 Anaesthetize the earthworm by placing it in MS 222 solution until it stops responding violently to manipulation. Decerebrate it by cutting off the anterior end behind the second segment. Rinse off the solution under the tap.

2 Locate the ventral surface (underside) and hold the worm firmly about 1 cm behind the clitellum (saddle). With a pair of small scissors, make a slit through the body wall about 2 cm long. Do not insert the point of the scissors too far, as the nerve cord and gut lie immediately beneath. If possible, leave the peritoneum, the thin membrane covering the cord, intact.

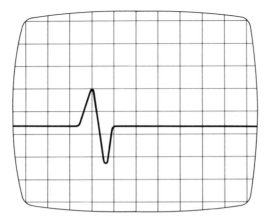

Figure 19
An action potential displayed on an oscilloscope screen.

3 Pin the worm, ventral surface uppermost, to the bottom of a metal dissecting dish, as shown in *figure 20*.

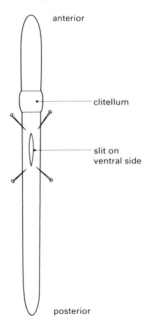

Figure 20
A worm opened up and pinned down.
Based on Roberts, M. B. V., Biology. A functional approach, *Students' manual, Nelson, 1974.*

4 Make a longitudinal slit in the peritoneum; this should release the coelomic fluid and expose the cord. Grasp the body wall with forceps and carefully cut the septa that lie beneath with small scissors or a sharp scalpel. The septa are the connective tissue partitions between adjacent segments and once they are cut the body wall can be pinned back on either side, as shown in *figure 21*.

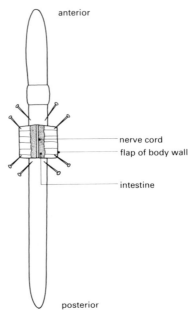

Figure 21
The body wall deflected and pinned back.
Based on Roberts, M. B. V., Biology. A functional approach, Students' manual, Nelson, 1974.

5 Keep the preparation moistened with Ringer's solution throughout the experiment. The cord must now be freed from the underlying intestine by lifting it gently with a seeker and cutting through any connective tissue beneath it. It is essential that the nerve cord should be clean and free of connective tissue.

6 Carefully lift the nerve cord onto the electrodes so that it is clear of the gut, but not stretched too much. If there is a film of water between the cord and the gut gently blow it away.

7 Earth the metal dish by connecting it to the earth input of the oscilloscope. An alternative earth lead to, for example, a water tap may be needed. If necessary run an earth lead from the preparation itself. The final arrangement is shown in *figure 22*.

8 Set the time base on the oscilloscope so that the spots just merge to

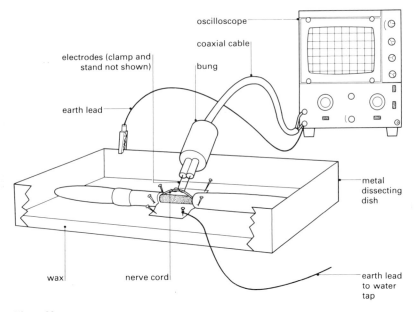

Figure 22
Recording impulses from the earthworm's ventral nerve cord. *Adapted from Roberts, M. B. V.*, Biology. A functional approach, Students' manual, *Nelson, 1974.*

form a continuous line; about $2\,\text{ms}\,\text{cm}^{-1}$ should be sufficient. The line must not be wavy or fuzzy; if it is, the earthing arrangements must be checked.

9   Set the gain (volts cm^{-1}) to the most sensitive setting, e.g. $10\,\text{mV}\,\text{cm}^{-1}$ to $50\,\text{mV}\,\text{cm}^{-1}$. If the oscilloscope will not reach this sensitivity, a pre-amplifier can be used to boost the signal.

10   Test the preparation by touching the anterior end of the worm with the point of a wooden-handled seeker. Action potentials should be visible on the screen. It is best to work in pairs so that one person can stimulate the worm while the other examines the potentials.

11   Use a pair of dividers to measure the exact height and width of the action potentials. Increasing the time base to $1\,\text{ms}\,\text{cm}^{-1}$ may help. From the known settings on the oscilloscope calculate the magnitude in millivolts, and the duration in milliseconds, of the potentials.

12   Repeat step **11** for potentials produced by touching the posterior end of the worm.

13   Investigate the effects of stimuli of different strengths by touching the worm very lightly and then more sharply. Record the numbers and sizes of the action potentials produced.

14   Investigate the effects of repetitive stimulation by touching the worm repeatedly at intervals of 1 to 5 seconds. Record results as before.

15 Additional investigations could include:
1 Comparing action potentials produced by stimulating the middle of the worm with those produced before.
2 Stimulating both ends of the worm simultaneously.
3 Comparing the effects of different stimuli, such as light, vibration, change in pH, air currents, chemicals.

Questions

a *How do the potentials produced by stimulating the two ends of the worm compare? Refer to magnitude (in mV) and duration (in ms) in your answer. Make scale drawings of the potentials from your measurements.*

b *What is the effect on the potentials of varying the strength of the stimulus? Can you give an explanation for this result?*

c *What is the effect of repetitive stimulation on the potentials? What possible explanations are there for this result? What experiments could be performed to test these ideas?*

d *Discuss the results and implications of any further investigations that you made.*

CHAPTER 12

THE RESPONSE TO STIMULI

INVESTIGATION
12A The response of *Tribolium* to humidity

(*Study guide* 12.2 'Animal responses'.)

Tribolium is a small beetle, a pest of stored food products, which attacks cereals and cereal products. *Tribolium confusum* is probably the most common flour beetle in mills and bakeries (*figure 23*). The other species commonly used in the laboratory is *Tribolium castaneum* (the red rust flour beetle). This requires a higher temperature than *T. confusum* to complete its life cycle and is therefore not such an important pest in this country.

Figure 23
An adult *Tribolium confusum*, dorsal view.
Courtesy of the Trustees of the British Museum (Natural History).

In this investigation you will be using a choice chamber (see *figure 24* overleaf). The upper part has a layer of muslin over the open base, held in position by an elastic band. The sections in the lower part can be filled with different reagents in order to produce two contrasting environments in the upper part. Animals can be inserted through the openings in the roof.

Procedure A: The orientation of Tribolium *in a humidity gradient*
1 Select a suitable position on a bench which is out of direct sunlight and not near a heat source. Do not jar the bench when you are doing this experiment.
2 Place the choice chamber on a piece of light-proof material, so that half of it can be folded over to cover the chamber.

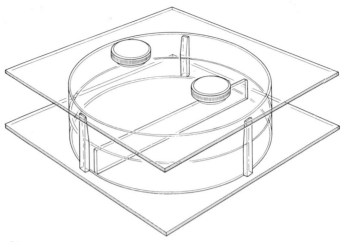

Figure 24
A choice chamber.

3　Arrange the chamber with the end of the central bar, in the lower part, facing you, so that the chamber is divided into right and left halves. Mark the position of the chamber on the bench with chalk.

4　Take the upper part and fix fine mesh material across the open base, using an elastic band. Pull the material taut to remove any creases and fit the two parts of the chamber securely together.

5　Collect ten *Tribolium* beetles in a Petri dish, using a pooter or small artist's brush. Drop five beetles through the right hole and five through the left hole. Seal both holes and cover the apparatus with light-proof material to exclude all light.

6　After ten minutes remove the cover and record the number of animals in each half of the chamber.

7　Rotate the upper part containing the animals through 180° and repeat your observations after a further ten minutes. Collect similar data from as many choice chambers as possible in order to make a statistical analysis.

8　Remove the upper part containing the animals and place it on a sheet of paper. Fill one half of the lower part with water or with a sponge soaked in water and the other half with a drying agent, such as anhydrous calcium chloride.

9　Replace the upper part and check the humidity by inserting a piece of cobalt chloride or cobalt thiocyanate paper through the holes in each half. Both types of paper are blue in a dry atmosphere and pink in a humid one.

10　Cover the chamber to exclude light as before and record the distribution of the animals after ten minutes.

---

11   Rotate the upper part through 180°, check that the humidity gradient becomes re-established, and repeat your observations after a further ten minutes. Pool the class results as before.

12   Analyse the class results using the χ^2 test. (See *Mathematics for biologists*.)

Questions

a   *State the null hypothesis used in the analysis of the data.*

b   *What was the purpose of the first series of observations on the beetles (stages 6 and 7)?*

c   *Why were the upper parts of the chambers rotated through 180° (stages 7 and 11)?*

d   *Comment fully on the results of the analysis (stage 12).*

Procedure B: The nature of the response of Tribolium *to humidity*

1   In order to find out what type of response is shown by the beetles, you need to follow the movements of individual beetles. Collect ten unused *Tribolium* beetles and set up a choice chamber with water and a drying agent as before (stage **8**). Leave it for five minutes to equilibrate.

2   Place one beetle in the chamber, start a stopclock, and map the beetle's movements by tracing on the lid with a felt pen or wax pencil. Put a cross on the trace at ten second intervals. Continue for 1–2 minutes.

3   Trace the map onto paper and remove the beetle.

4   Repeat stages **2** and **3** with the other beetles, ensuring that none is used more than once, and that they are started alternately on the dry and humid sides.

5   For each environment (dry and humid) calculate (a) the mean velocity and (b) the mean number of turns made per metre travelled.

Questions

e   *Does the velocity of the beetles vary with humidity? If so, explain how this could bring about the observed orientation in a humidity gradient.*

f   *Does the rate of turning of the beetles vary with humidity? How could this bring about the orientation in a humidity gradient?*

g   *What do the results suggest about the direction of the beetles' movement?*

(Remember to return the beetles to their culture after use.)

INVESTIGATION
12B The response of *Calliphora* larvae to light

(*Study guide* 12.2 'Animal responses'.)

This investigation is designed to study the type of response which an animal exhibits towards light. *Calliphora* larvae, the maggots of the bluebottle, are convenient animals to use for this purpose. Fully grown larvae, just before pupation, are the most suitable (*figure 25*).

Figure 25
A larva of *Calliphora* sp.

Procedure

1 Take a piece of black card, about 30 cm square, and fix it to the bench with adhesive tape, ensuring that it is as flat as possible. Using two retort stands and clamps, support a sheet of glass in a horizontal position about 5 cm above the card. Place a lamp just above the edge of the card halfway along one side.

2 Black out the laboratory and carry out the rest of the investigation with only the experimental lamp on. Take precautions to cut out the light from other sets of apparatus.

3 Collect ten *Calliphora* larvae in a Petri dish. Release one animal near the centre of the card and follow its movements by marking the glass with a felt pen or wax pencil.

4 Repeat this procedure several times with the same animal, starting it pointing in a different direction each time. Trace the maps onto paper, clean the glass sheet, and repeat with other individuals until you are reasonably sure of their response to light.

5 Place a second lamp at 90° to the first, about halfway along an adjacent side. Ensure that the bulbs are of the same strength, testing with a light meter if necessary. Draw a faint mark on the card to indicate the position of the second beam.

6 With only the first lamp alight, place a larva on the edge of the card just in front of the lamp. Reject any animals that do not respond as

before. Follow the path of the animal and when it reaches the position of the second beam, switch the second lamp on and the first lamp off. Repeat this several times with the same animal and with other individuals.

7 Repeat stage **6**, but this time leave the first lamp on when you switch on the second. Record the results as before.

8 Replace the second lamp with a more powerful one or alter the positions of the lamps to reduce the intensity of the first one. Repeat as in stage **7**.

9 With a ruler and protractor, measure the angles of the animals' paths in relation to the direction of the first beam. Return them to their culture.

Questions

a *Is the larva's response to light directional or random?*

b *If there was directional movement, was it towards the light (photopositive) or away from the light (photonegative)?*

c *What other factors could have affected the response and how could you try to eliminate their effects?*

d *How can you account for the fact that different individuals may not all respond in the same way?*

e *How do the animals respond to the second light beam when the first is switched off?*

f *How do the animals respond to the second beam when the first is left on?*

g *What effect does an increase in intensity of the second beam have on the orientation of the larvae?*

h *What is the importance of these responses to light in the normal life of the larvae?*

INVESTIGATION

12C Short-term changes in responsiveness

(*Study guide* 12.2 'Animal responses'.)

An animal's responsiveness to a particular stimulus can often show changes in the short term. In these two investigations you will study changes in the earthworm's responsiveness to the stimulus of touch.

Part I Waning of the response due to repeated stimulation

Procedure

1 Moisten the inside of a dish with several drops of water, place the worm in it, and leave for at least 15 minutes under dim illumination.
2 Stimulate the worm by touching it for one second with a seeker, and note the response.
3 Investigate the effect of repeated stimulation by giving the worm stimuli at fixed intervals of between 10 and 20 seconds, keeping the duration and strength of the stimuli as similar as possible. It is important that the worm is stimulated in about the same place each time and that the light intensity is at a low and unvarying level. Observe and record the worm's response each time. Continue until the worm has ceased to respond to five stimuli in succession.
4 Once the worm has become unresponsive to a particular stimulus, continue to stimulate in a *new* position. Record all responses as before.
5 Allow an unresponsive worm a period of 10–20 minutes without stimulation. Stimulate it again and if the response has recovered continue as in stage **3**. If recovery does not take place, longer rest periods should be given.
6 If possible, repeat the test procedures on subsequent days with the same animals. Return them to their habitat.

Questions

a *1 How does the earthworm respond to a single stimulus of touch?*
2 How does the response change with repeated stimulation?
3 How many stimuli were given before the worm became unresponsive?

b *1 What is the effect of stimulating an unresponsive worm in a different place?*
2 What does this indicate about the possible mechanisms underlying the waning of the response?

c *How does the rate of waning to the new stimulus compare with the rate of waning to the first stimulus?*

d *1 Does recovery of the response occur with time, and if so, is this recovery (i) rapid, and (ii) complete?*
2 If recovery takes place in distinct phases, what might this indicate about the initial waning?

Part II   The effect of dual stimulation on responsiveness

Procedure

1   The contact of the worm's body with the walls of its burrow is a source of stimulation that operates when specific stimuli are applied. Repeat the procedure as in stage **3** above using a worm whose body wall is in contact with a rough surface. As much as possible of the body wall should be touching the surface. Stimulate as before, recording the responses until the worm ceases to respond.

2   If there is time light intensity could be varied to see if this acts to modify the worm's responsiveness.

Questions

a   *Does the degree of contact between the worm and the rough surface affect the response of the worm to touch?*

b   *In what way does light intensity affect the earthworm's responsiveness?*

c   *Review your results and suggest how short-term changes in responsiveness, such as the ones you have investigated, might be to the advantage of the organism.*

INVESTIGATION
12D   The tropic responses of coleoptiles and radicles

(*Study guide* 12.3 'A plant's response to light' and 12.4 'A plant's response to gravity'.)

A tropism is a response in plants by growth curvature, where the direction of response is related to the direction of the stimulus. This investigation is concerned with the response to light (phototropism) and the response to gravity (gravitropism).

Procedure A: The effect of unidirectional light on a coleoptile

1   Monocotyledonous plants, such as grasses and cereals, have their first leaves enclosed in a cylindrical sheath called a coleoptile, which because of its simple, regular shape provides ideal material for the study of phototropism. Take a germinating wheat grain which has an intact coleoptile (i.e. with no leaves protruding) and set it in a mount as shown in *figure 26* overleaf.

2   Fit a graticule (linear scale) into the eyepiece of a microscope. Bend wire into the shape shown in *figure 27* overleaf and push this gently into position on top of the graticule to hold it firmly in place.

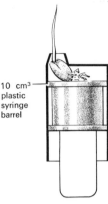

10 cm³
plastic
syringe
barrel

Figure 26
A mount for holding a coleoptile on a microscope stage.

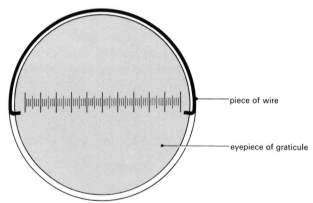

piece of wire

eyepiece of graticule

Figure 27
A means of fixing a graticule.

3   Arrange the microscope horizontally so that light, either from the windows or from a lamp, strikes it from one side only. Fix the coleoptile, in its mount, vertically to the microscope stage, using rubber bands or Blu-tack.

4   Focus the low power objective ($\times 3$ to $\times 5$) on the coleoptile and adjust the mounting device until the coleoptile tip is lined up with the mid-point of the graticule scale. Set the scale horizontally as shown in *figure 28*.

5   Note the time and start a stopclock. After two minutes note the position of the tip of the coleoptile against the scale. Make slight adjustments of the graticule, if necessary, so that the coleoptile tip appears to move along the scale rather than through it. Bear in mind, that when viewed through a monocular microscope, the direction of

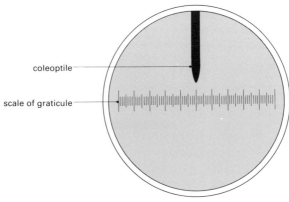

Figure 28
A coleoptile tip and a graticule scale.

movement appears reversed.

6 Take further readings of the coleoptile position at 1–2 minute intervals until a definite and regular movement has been recorded. This should take about 10 minutes.

7 Note the time and rotate the mounted coleoptile through 180° on its vertical axis. Reset the coleoptile tip on the mid-point of the graticule scale and take readings as before for a further 30 minutes.

8 Plot the successive positions of the coleoptile on graph paper.

9 If possible, leave the coleoptile set up for an hour or more and then remove it and draw its final shape in relation to the position of the light source.

Procedure B: The effect of gravity on a radicle

1 Take a mung bean or pea with a fairly straight radicle, 1–2 cm long, and fit it into the mount shown in *figure 26*. To prevent desiccation make sure that the radicle is kept moist.

2 Fit a graticule into a microscope eyepiece as shown in *figure 27*. Arrange the microscope horizontally and fix the mounted radicle, also horizontally, to the microscope stage.

3 Using the low power objective, set the graticule scale vertically, and adjust the radicle until its tip is lined up with the mid-point of the scale, as shown in *figure 29* overleaf.

4 Start a stopclock and record the position of the radicle tip at 1–2 minute intervals for about 10 minutes.

5 Note the time and rotate the mounted radicle through 180° on its horizontal axis. Reset the radicle tip on the mid-point of the scale and continue to take readings for a further 30 minutes.

6 Plot the readings on graph paper and, if possible, draw the shape of the radicle after an hour or more.

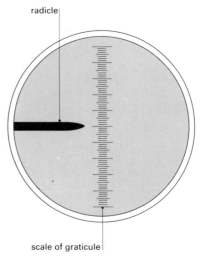

Figure 29
A radicle tip and a graticule scale.

Questions

a   *How do the coleoptile and radicle respond to the stimuli of light and gravity respectively?*

b   *When the plant organs are rotated through 180°, the direction of the stimulus was, in effect, reversed. Was the movement of the organ immediately reversed? If not, what was the period of delay?*

c   *Did the whole plant organ bend, or was there a definite region which responded?*

d   *What advantages would these responses bring to a plant growing in its natural habitat?*

e   *Suggest a hypothesis to account for your experimental results.*

INVESTIGATION

12E The nature and detection of the stimulus involved in phototropism

(*Study guide* 12.3 'A plant's response to light'.)

Coleoptiles exhibit positive phototropism, that is, they respond to the stimulus of light by bending towards it. However, white light is not a simple stimulus; it is a mixture of different colours, each colour having a different wavelength. One aim of this investigation is to discover which colours are important in phototropism. A second but equally important aim is to find out whether all parts of the coleoptile are sensitive to light, and, if not, which parts are sensitive.

Procedure

1 You are provided with several pots containing a number of coleoptiles that have been grown in the dark. It is vital that they are not exposed to any light prior to the experiment, though red light can be used in a darkened room.

2 Remove any coleoptiles that are not complete, that is, those in which the leaves have broken through the coleoptile.

3 Make a number of small aluminium caps to cover the top 3 mm of the coleoptiles. Place these over one-third of the coleoptiles in each pot (*figure 30*).

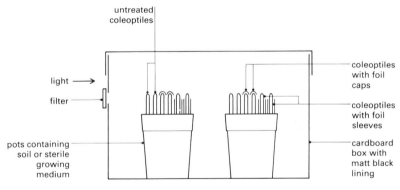

Figure 30
A light box containing coleoptiles of germinating wheat seedlings.

4 Make a number of aluminium sleeves to cover all but the top 3 mm of the coleoptiles. Place these around another third of the coleoptiles in each pot (*figure 30*).

5 Leave the remaining third of the coleoptiles untreated. Place the pots in light boxes as shown in *figure 30*, so that the coleoptiles receive unilateral light of a particular colour, depending on the filter used. Red

and blue light should be tested and suitable controls included. Other colours could also be tried.

6 Leave the boxes for 3–4 hours, after which the responses of the coleoptiles can be observed and recorded.

Questions

a *What controls did you use for this experiment?*

b *To which colours of light was a response shown?*

c *Elsewhere in the course you will read about the pigment known as phytochrome (Study guide II, Chapter 25). Do these results indicate that phytochrome is involved in the phototropic response of coleoptiles? Give your reasons.*

d *Which part of the coleoptile is sensitive to light?*

e *Which part of the coleoptile responds, that is, which part actually bends?*

f *Suggest hypotheses to explain your answers to questions d and e.*

g *What further experiments could be done to test your hypotheses?*

h *Why do you think that matt black surfaces are needed inside the box rather than white or silvered surfaces?*

i *Explain why red light can be used in a darkened room before the experiment.*

BEHAVIOUR

INVESTIGATION
13A **Turning behaviour in woodlice**

(*Study guide* 13.2 'Studying the causation of behaviour'.)

This investigation is concerned with the causal explanation of the behaviour shown by woodlice in a maze. You will be trying to find out why the animals turn in a particular direction when given a choice between turning left or right. One possibility is that the woodlouse's behaviour at a choice point is influenced by its earlier behaviour, especially the direction in which it has previously turned. The woodlice will be forced to turn through 90 degrees and the effect of this on the next free turn will be investigated.

Procedure

1   The woodlice are to be tested in the maze shown in *figure 31*. Half the animals should start from position A and half from position A'. The alley blocks can be placed to produce a forced turn, to the left or right, at point V. Free turns can be allowed at points W, X, Y, or Z. A bench lamp placed over the maze will keep the illumination relatively constant.

2   Place the first woodlouse at A, with the alley blocks in position as shown in *figure 31*, to allow the free turn at point W. Record the choice made at W as 'same direction' (S) or 'opposite direction' (O), compared with the direction of the forced turn.

3   Repeat, starting the second woodlouse at A', the third at A, and so on. Change the position of the alley block at V accordingly. Test each woodlouse once only and keep it for further experiments. Perform at least 40 trials.

4   By placing additional alley blocks at the appropriate places in the maze, the woodlice can be made to have their free turns at W, X, Y, or Z. For example to allow a free turn at X, the side arms at W must be blocked. In this way the distance between the forced turn and the free turn can be increased. Test 10 woodlice for each distance, starting alternately from A and A'. Record choices made as (S) or (O), as before. In addition, record the time taken for each woodlouse to move from point V to the free turn point.

5   Calculate the median times for the woodlice to travel distances WX, WY, and WZ in stage **4**.

6   Return the alley blocks to the original position shown in *figure 31*. The woodlice can be delayed at point D by applying an artist's brush to

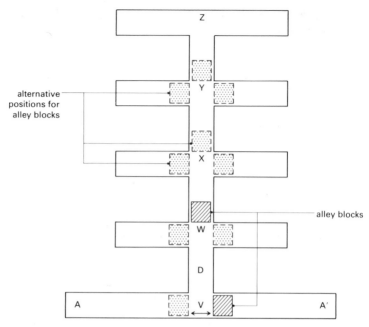

Figure 31
A woodlouse maze.

their dorsal surface. This has the effect of varying the time which
elapses between the forced and the free turn, while the distance
travelled is kept constant. Delay the woodlice for times equal to the
median times calculated in stage **5**. Test 10 woodlice for each period of
delay, including 10 which receive no delay. Record results as before.

7   Pool the class results and add up the total number of (S) and (O)
    choices made in each part of the investigation.

8   Use the χ^2 test to see whether the proportion of (S) and (O) choices is
    significantly different from that expected by chance (i.e. 50 per cent).

9   Draw graphs of the percentage (O) choices against distance between
    the forced and free turns and against delay period between the forced
    and free turns.

Questions

a   ***Do the results indicate that the direction of a forced turn influences
    the direction of a subsequent free turn?***

b   ***If your answer to question a is yes, is the influence in the same
    direction as, or the opposite direction to, that of the forced turn?***

c *Does the direction of a free turn vary with the distance between the forced and free turns?*

d *Compare the results of stages 4 and 6. Does the influence of one turn on the next vary with the distance between the turns or with the time which elapses between the turns?*

INVESTIGATION
13B The reproductive behaviour of the three-spined stickleback

(*Study guide* 13.2 'Studying the causation of behaviour'.)

In all forms of social behaviour, communication is involved between the participants. The classical work in analysing the reproductive behaviour of an animal was carried out by Tinbergen on the common three-spined stickleback, *Gasterosteus aculeatus*, which can be found in nearly every pond. With patience you can observe stickleback behaviour in an aquarium during early summer. Failing that, you can make use of a film, or observe other animals, such as zebra finches or *Drosophila*.

Procedure
1 Set up an aquarium with suitable gravel and water plants. Provide filamentous algae and string as nest-building materials. Divide the aquarium into two halves with an opaque glass partition and place a male stickleback, in breeding condition, in each half. Observe and record the activities of the fish over the next few days. Most males will show nest-building behaviour after being isolated. Describe the behaviour patterns involved as accurately as possible.
2 Replace the opaque glass partition with a transparent one. Observe and record any interactions which take place over the next 10 to 15 minutes.
3 Remove the partition completely and repeat your observations.
4 Replace the original partition and present one fish with an image of itself by inserting a small mirror in the water. Record the responses.
5 Offer one fish a model emphasizing a characteristic feature of male sticklebacks, such as red coloration, and record any responses made. Do not perform too many experiments with the same fish on any one day.
6 Add a gravid female fish (one swollen with eggs) to the territory of one of the males and record any interactions that take place. Remove the female after 15 minutes.
7 When a nest has been successfully completed, reintroduce the female fish and carefully record all activities and the time they take. If the female is attacked vigorously, remove her.

8 If courtship is successful and leads to the laying of eggs, remove the female and continue your observations of the male fish during the time the eggs are developing and after they hatch.

9 Offer the other male fish a model emphasizing the shape of a female and record any responses made. Compare the effects of models of different shape, colour, degree of motion, and so on.

Questions

a *What processes are involved in building a nest? What discrimination does the male show during nest-building?*

b *In what ways do the appearance and behaviour of a male change when*
   1   it sees another male, or
   2   another male enters its territory, or
   3   it sees its own reflection?

c *What feature of a male seems to be most significant in releasing aggressive behaviour?*

d *What advantages are gained by the male and the species in general as a result of the establishment and defence of territories?*

e *Describe the movement of a breeding male towards a receptive female. Summarize the responses of courtship as shown on the left side of figure 32.*

f *What stimuli appear to be important in releasing the various courtship responses, for example, the zig-zag dance of the male or egg laying by the female?*

g *Did any of the behaviour patterns of the fish take place out of context? Describe the occurrence of these apparently irrelevant or displacement activities.*

INVESTIGATION
13C   Sex pheromones in the Mediterranean flour moth

(*Study guide* 13.2 'Studying the causation of behaviour'.)

A pheromone is a chemical substance released into the environment by an organism that influences the behaviour or development of other individuals of the same species. For example, many species of ants follow chemical trails that have been produced by workers once they have found food. The trail is reinforced by the secretions of other workers and is likely to be maintained until all the food has been carried to the nest.

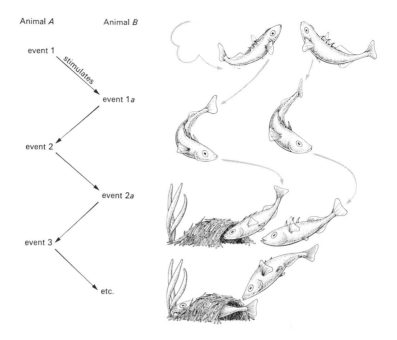

Figure 32
The sequence of courtship of the three-spined stickleback. *Partly based on Tinbergen, N., The study of instinct, Oxford University Press, 1951.*

This investigation involves another important function of phero-mones, that is their use as sexual attractants. The organisms used are Mediterranean flour moths, *Ephestia kühniella*, which are easily handled as they are practically flightless.

Procedure
1 You are provided with two specimen tubes, one containing a female moth and the other several males. The sexes have been separated from each other for at least a day beforehand.
2 Place a gauze or small piece of nylon over the mouth of the tube containing the female. Place the tube, upright, on the bottom of an aquarium tank as shown in *figure 33* overleaf.
3 Set up a similar tube as a control, containing a male moth. Position the two tubes at either end of the tank.
4 Release several males into the tank and replace the glass lid.
5 Observe the behaviour of the males over the next five minutes. Record the number of males that come into contact with the two specimen tubes. Record any differences in behaviour or appearance of the moths inside the specimen tubes.

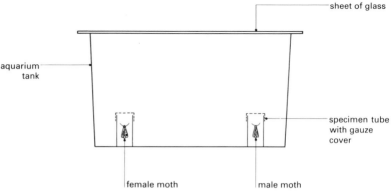

Figure 33
An aquarium tank containing *Ephestia* moths.

6   Recapture the males and keep them in a specimen tube for further
    experiments.
7   Remove the female moth and kill it by decapitation with a sharp
    scalpel.
8   Carefully remove the fine abdomen tip by making a transverse cut
    2–3 mm from the end. Place the tip in the centre of a watch-glass in the
    bottom of the aquarium tank.
9   Place the rest of the body of the female in the centre of another
    watchglass as a control. Position the two watch-glasses at either end of
    the tank as shown in *figure 34*.

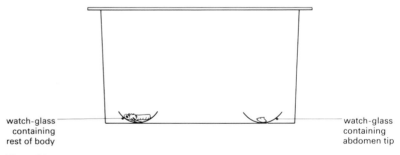

Figure 34
An aquarium tank containing a dissected *Ephestia* female.

10 Release the males into the tank as before. Record any similarity to
    their behaviour in the first experiment. Count the number of males
    that come into contact with the watch-glasses.
11 Recapture the males, remove the tip and discard the rest of the body of
    the female moth. Grind the tip up in several drops of ether and place a
    drop of the extract in the centre of a clean watch-glass.

12 Place a drop of pure ether in the centre of another watch-glass as a control and position both watch-glasses at either end of the tank.

13 Release the males into the tank and record the results as before.

14 It is likely that the response of the males to the extract will wane, that is, decrease with continued exposure to the stimulus. If this occurs, see if the response can be restored by the addition of a fresh drop of extract.

Questions

a *How do the male* **Ephestia** *moths respond to the presence of a female?*

b *To what types of stimuli could the male moths be responding?*

c *How far do the results of stage 10 help you to decide what type of stimulus the moths are responding to?*

d *What do the results of stage 13 suggest about the nature of the stimulus?*

e *Suggest reasons for the waning of the males' response to the extract.*

f *What further experiments should be done to investigate the mechanisms underlying this behaviour?*

INVESTIGATION

13D Associative learning in mammals

(*Study guide* 13.2 'Studying the causation of behaviour'.)

Many animals modify their behaviour as a result of previous experience. Changes of this sort are generally known as learning and in this investigation you will study one type of learning where an animal develops an association between a new behaviour pattern and the consequences of performing the new behaviour.

Procedure

1 Any small mammals are suitable, the minimum number being two. They must be kept in separate, identical home cages and should be accustomed to being handled. One animal will be the experimental subject, the other will act as a companion. Throughout this investigation take care not to cause any stress to the animals.

2 The animals will be tested in a double-choice learning box as shown in

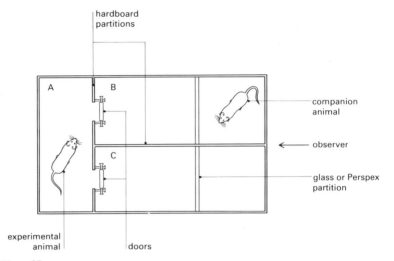

Figure 35
Plan of a double-choice learning box.

figure 35. Place the companion animal in chamber B as shown, providing food and bedding material as necessary. This animal should remain here throughout the trials.

3 Place the experimental animal in chamber A and start a stopclock. Observe discretely from the position shown. It is vital that the animal is not disturbed by stimuli other than those inside the box.

4 Record the chamber chosen (B or C) and the time taken for the animal to enter it. Allow the animal 10 seconds to explore its new surroundings before removing it.

5 Return the animal to chamber A and repeat the procedure, noting on each occasion which chamber is entered. If the animal takes longer than five minutes to make a choice, remove it from chamber A and start the trial again.

6 Aim to complete at least twenty trials with the same animal, and then return it to its home cage.

7 Perform a control experiment with a second animal, in which chamber B contains no companion. Record your results as before.

8 If possible repeat the trials and controls at intervals of three or more days.

9 Draw a graph of the results for each animal as shown in *figure 36.* Calculate the mean time taken for the animal to make a choice (of either chamber B or C) for the first ten trials and for the last ten trials.

10 If other animals are available the experiments can be replicated and

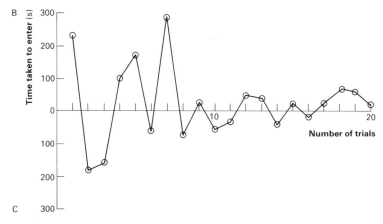

Figure 36
A learning graph.

the results pooled. The χ^2 test could be applied to see if the number of entries into chamber B is significantly different from that expected by chance.

Questions

a *Compare results for the experimental and the control animals. What do the results indicate as far as the time taken to make a choice is concerned?*

b *What do the results indicate as far as choices for chamber B are concerned?*

c *Explain as fully as you can the differences between the experimental and control results.*

d *Learning of this type is sometimes known as operant or instrumental conditioning. To which features of the experiment would you apply the following terms?*
1 Conditioned stimulus.
2 Conditioned response.
3 Reinforcing stimulus.
4 Unconditioned response.

e *Explain how this type of learning might be of value in the wild. Can you think of any advantage to the animal in continuing to make occasional entries into chamber C?*

INVESTIGATION
13E   Social order in hens

(*Study guide* 13.4 'Studying the functions of behaviour'.)

Hens, like many other animals, develop a social order or hierarchy after being placed together for some time. In this investigation you will observe the behaviour of a small group of hens, at first hand if they are available, or by watching a film or film loop. Alternatively you can perform similar work with mice or other small mammals. Again, take care to avoid causing any stress to the animals.

Procedure

1   You are provided with four or five animals that have been marked for identification and housed separately for at least a week beforehand. Place one of the animals in an observation pen and observe its behaviour for 5–10 minutes. Record any characteristic movements or postures.

2   Place a second animal in the pen and observe for a similar period of time. Note particularly any behaviour that arises from interactions between the two animals, such as one bird pecking the other. Try to decide which behaviour patterns indicate dominance and submission.

3   Draw a large grid as shown in *figure 37*.

4   Place the remaining animals in the pen and observe their behaviour as before. Record the results of any interactions between the animals in the grid, that is, which animal is dominant and which is submissive. It may help to work as a group, with each student concentrating on a particular animal. Do not confuse dominant behaviour with alertness to the human observer, or submissiveness with food searching or resting. Continue observations until a clear pattern emerges.

5   Repeat these observations after the animals have been together for at least one week. Record your results on a similar grid. In addition, throw a crumb of food into the pen from time to time and record in a separate grid which animal gets the food. It will help if the animals have not been recently fed.

6   If you have time you could investigate the short- and long-term effects of adding strange animals to the existing group.

Questions

a   **Which behaviour patterns are associated with the following:**
   1   **threat;**
   2   **attack;**
   3   **appeasement;**
   4   **escape?**

b   *Which patterns are shown by dominant animals and which by submissive ones?*

c   *How far is it possible to work out the social order or hierarchy of the animals?*

d   *Is the social order maintained in an established group?*

e   *Is the social order correlated with success in feeding when food is scarce?*

f   *What do you think may be the function of an established order in a group of animals in the wild?*

		Animal exhibiting dominant behaviour				Total 'losses'
		1	2	3	4	
Animal exhibiting submissive behaviour	1	—				
	2		—			
	3			—		
	4				—	
Total 'wins'						

Figure 37
The outcome of social encounters in a group of four animals.

THE HUMAN BRAIN AND THE MIND

INVESTIGATION
14A Sensorimotor skills in humans

(*Study guide* 14.1 'The human brain'.)

Part 1

When we draw, we use a form of skill called sensorimotor. That is a skill where muscular control is altered by feedback through the sensory system. In mirror drawing our normal co-ordination between hand and eye is useless and we have to learn a new pattern.

Procedure

1   You should work in pairs for this investigation. One partner should act as the experimental subject while the other records, and then you should exchange roles.

2   Set up the apparatus as shown in *figure 38*. The mirror should be about 50 cm away from the subject. The hardboard sheet must be high enough to enable the subject to write with a pencil beneath it, but not

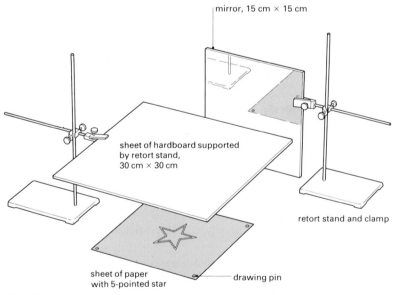

mirror, 15 cm × 15 cm

sheet of hardboard supported by retort stand, 30 cm × 30 cm

retort stand and clamp

sheet of paper with 5-pointed star

drawing pin

Figure 38
The apparatus for mirror drawing.

so high as to block the mirror. Place the five-pointed star figure, which has a double outline, beneath the sheet.

3   At a given signal, the subject should trace a pencil line around the star and in between the two lines. He should do this as quickly as possible, trying to avoid crossing either line with the pencil. He must not touch the bench with any part of his body.

4   Repeat this at least 15 times, using a fresh star figure each time. Fix the figure in the same place and start tracing from the same point.

5   For each trial record the time taken and the number of errors, that is, the number of times the pencil line crossed one of the printed lines of the star figure.

6   Plot graphs of the time taken and the number of errors against the number of the trial.

Questions

a   **What do the graphs indicate about the speed at which you can learn this type of skill?**

b   **How much variation in the rate of learning is there in the class?**

c   **How could you find out if the skill you have gained on this particular drawing makes it easier to produce a different mirror drawing?**

Part 2

We have seen that learning is involved in the development of a sensorimotor skill. Learning may be described as an adaptive change in behaviour resulting from experience. The importance of experience can be shown by investigating the effect of knowing the result of a previous performance on subsequent performances.

Procedure A

1   The subject should try to draw freehand a straight line as near 7.5 cm long as possible. The piece of paper should then be turned face down on the bench and a line drawn on a second sheet as before. Continue this for 20 trials.

2   The 21st sheet should be handed to the partner who measures it to the nearest 0.5 cm and then tells the subject the result. Only then should the subject draw a line on the next sheet. Perform 20 trials.

3   Measure the lengths of the lines in both series of trials and draw a graph of the results against the number of the trial. Include both series on the same graph.

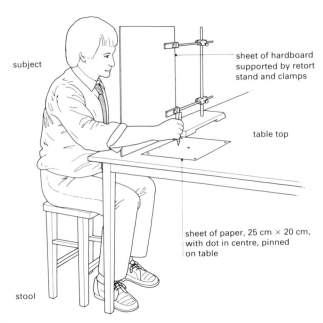

subject

sheet of hardboard supported by retort stand and clamps

table top

sheet of paper, 25 cm × 20 cm, with dot in centre, pinned on table

stool

Figure 39
The dot learning test.

Procedure B

1 Set up the apparatus as shown in *figure 39*. The subject must be able to mark the sheet of paper around the dot, but be unable to see it.

2 The subject tries to mark the paper as near to the dot as possible, without touching the bench. After each attempt the partner should measure the distance between the dot and the pencil mark and record it, without informing the subject. Continue for 20 trials.

3 Perform a second series of trials, but this time the subject should be informed of the result of the previous trial, e.g. '5 cm north-west of the dot'.

4 Plot the results of the two series of trials against the number of the trial on the same graph.

Questions

a *Did these experiments indicate that knowledge of the results enhanced subsequent performances?*

b *What other information must be available to the brain, in all these experiments, if performance is to be improved?*

c *How might performance be affected if you ran a series of 50 trials or more without a break?*

d *On what occasions are the results of these investigations used in everyday life?*

INVESTIGATION
14B Perception

(*Study guide* 14.2 'The mind and consciousness'.)

The eyes and the brain work together to convert the impressions that are received from the outside world into an image that can be easily recognized and understood. But if the cues provided by an image are ambiguous the brain may be misled and perceive the image in two quite different ways. This is the way visual illusions arise.

Procedure

1   Examine the visual illusions shown in *figure 40* overleaf.
2   Work in pairs to investigate the familiar Müller-Lyer illusion. Both partners should draw a reference line exactly 20 mm long.
3   Draw a series of accurately measured lines ranging from 15–25 mm in length. One partner should add 'lengthening' arrows to the lines, i.e. $\rangle\!\!-\!\!-\!\!\langle$, while the other partner adds 'shortening' arrows, i.e. $\langle\!\!-\!\!-\!\!\rangle$.
4   Swop papers and estimate the lengths of the arrowed lines by comparing them with the reference line. Write your estimates, to the nearest mm, beside the lines.
5   Plot the estimated lengths against the actual lengths for each set of lines, on the same graph.

Questions

a *Describe how the brain is confused or misled by the illusions shown in* **figure 40.**

b *What additional cues might resolve the ambiguities in the drawings?*

c *What do your results suggest about the effect of adding arrows to the measured lines?*

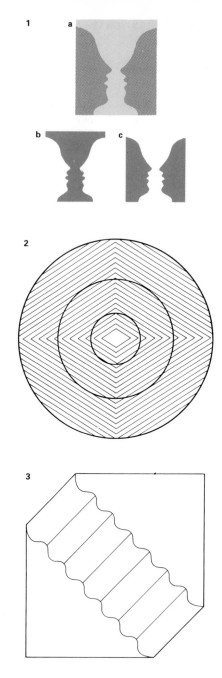

Figure 40
Visual illusions.

INVESTIGATION
14C Memory

(*Study guide* 14.2 'The mind and consciousness': 'Memory and learning'.)

Memory is the storage and recall of information by the brain. Simple tests can be used to investigate short-term memory and in particular the differing abilities of people to remember different things. Before you start, make sure you understand the different functions of the two cerebral hemispheres.

Procedure

1 You are shown a tray containing about 15 objects, for one minute.
2 After the tray is removed, list those objects that you can remember.
3 Repeat with trays of names, numbers, and famous faces.
4 Tabulate the scores on the four tests for each member of the class.
5 Perform a fifth test with a second tray containing the names of the famous faces and add the results to the table.
6 You are shown an abstract design which you should copy. Later in the lesson you will be asked to remember and re-draw it. Your performance can be scored and the results added to the table.
7 A 100-word story is read to you. You will be asked to recall it verbally immediately afterwards and again later. Add the scores to the table.
8 Divide the 100-word story into four parts. Score both of your versions of each part separately.

Questions

a *Did you find that some people scored better on some tests compared with others? Was there a difference in the scoring between males and females?*

b *Which tests indicate dominant hemisphere functioning and which indicate non-dominant hemisphere functioning?*

c *Can you explain your answer to question a in terms of the efficiency of the two cerebral hemispheres?*

d *Account for the results of the test using the second tray of names (stage 5).*

e *Is there any evidence from stage 8 that you remembered some parts of the story better than others?*

Tests of numerical ability

1

Find the concept which gives the missing number in the octagon.

2

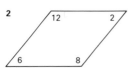

Use the first two completed examples to form a concept and then fill in the missing number in the pentagon.

Tests of visio–spatial ability

3

  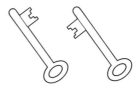

Choose the shape which is the odd one out.

4

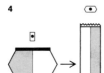 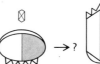

Follow the sequence and then choose the corresponding shape from one of the four possibilities given alongside.

Figure 41
Intelligence tests.

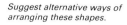
Suggest alternative ways of
arranging these shapes.

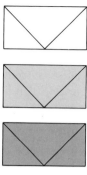

Figure 42
A test for conventional and creative thinking.

INVESTIGATION
14D Intelligence

(*Study guide* 14.2 'The mind and consciousness': 'Intelligence'.)

Intelligence encompasses a wide variety of mental processes and is therefore very difficult to measure. It is usual to test particular skills that seem to be reliable parameters of intelligence, such as numerical ability, visio-spatial skills, and verbal skills.

Procedure

1   *Figure 41* shows a number of intelligence tests. Complete them to the best of your ability and then score your performance. Your teacher will provide the solutions.
2   Compare the class results for each test.
3   Creative (lateral) thinking is a process not usually measured by standard I.Q. tests. Write down how many uses you can think of for a brick.
4   Complete the test illustrated in *figure 42* which is designed to discriminate between conventional and creative thinking.

Questions

a   *Do the results of the intelligence tests follow any pattern? Is there a difference in the performance of the class between boys and girls?*

b   *Do you have any criticisms of the tests?*

c   *How does conventional thinking differ from creative thinking as shown by* figure 42?

INVESTIGATION
14E Emotion

(*Study guide* 14.2 'The mind and consciousness': 'Emotion and feeling'.)

Emotion is characterized by the arousal of brain and body. Three parameters can be used to measure the state of bodily arousal: pulse-rate, skin response, and pupillary size.

Procedure

1 Before carrying out tests you need to establish a 'base-line' for each parameter, under normal conditions. It is easiest to work in pairs.

2 Find your partner's pulse-rate by counting the number of pulses at the wrist for 15 seconds and then multiplying by four. Repeat three times and take the average.

3 Hold your partner's hand and rate it for moistness on a 0–5 scale, where 0 is dry and 5 is very moist; and then for temperature, where 0 is warm and 5 is cold.

4 Ask your partner to look at a blank wall from a particular position in the room and rate his pupil size on a 0–5 scale.

5 Having established the basic levels, measure changes in arousal during the rest of the lesson. Your teacher will provide suitable stimuli.

6 Keep a day's record of your pulse-rate, taking it any time you feel upset, excited, or aroused. See *figure 43*.

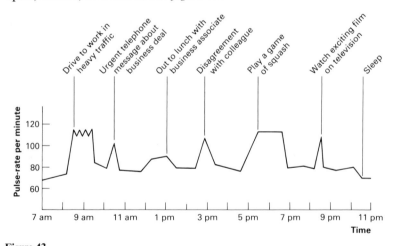

Figure 43
The way in which a person's pulse-rate may change during the day, and reasons for the changes.

Questions

a *To what extent are the three parameters (the pulse-rate, the skin response, and the pupillary size) correlated in different people?*

b *How do the three parameters change during bodily arousal?*

c *What mechanisms might underlie the observed changes?*

INVESTIGATION

14F Sleep

(*Study guide* 14.2 'The mind and consciousness': 'Sleep'.)

Sleep seems to be an active state that follows a pattern of variable rhythms throughout the night. Our conscious, daytime state also seems to follow a similar pattern. Before you start this investigation, make sure you know about the different phases of sleep and the different theories of dreaming.

Procedure

1 Keep a record of which phase of sleep, slow wave or REM, you awake from in the morning.
2 Keep a dream record for a week. Try to remember to do this as soon as you wake up.
3 Take your body temperature as often as possible during the day, e.g. every 1–2 hours. Plot a graph to show how your temperature varies and compare it with those of others in the class.

Questions

a *Which phase of sleep do you usually wake from in the morning?*

b *Which theory of dreaming seems to fit your own experience?*

c *Do the graphs of body temperature show any common pattern? If so, does this relate to the time when individuals feel at their best or function most efficiently?*

INVESTIGATION

14G Extra-sensory perception and psychokinesis

(*Study guide* 14.2 'The mind and consciousness': 'Changing the way we see the world'.)

Simple ESP experiments can be conducted with a set of cards bearing five geometrical designs: a cross, a star, three wavy lines, a circle, and a

Chapter 14 The human brain and the mind   59

square. There are five cards of each design, so the probability of a particular card being guessed correctly is one in five.

Procedure

1 *Clairvoyance.* Place the cards face down on a table and try to guess their order. Record your guesses by writing down the symbols. Turn the cards over, one by one, and add up the number of correct guesses.

2 *Telepathy.* One person turns the cards over one by one, looking at the symbols, while the others record their guesses as before. Find the number of correct guesses after all the cards have been dealt.

3 *Psychokinesis.* Throw a die 30 times and record the frequency with which each face comes up. Repeat this, with the class 'willing' a particular face to come up and record the results as before. If there seems to be an effect, repeat this on your own at home.

4 The results you have obtained in these experiments could be analysed using the χ^2 test to see if they are significantly different from those expected by chance.

Questions

a **What do your results suggest about the existence of ESP and psychokinesis?**

b **What is the difference between stage 1 and stage 2?**

c **If some people seem to be particularly good at influencing the die at home, can they repeat their success in class? If not, why do you think they fail?**